How Heavy? How Much?

by Jared Adams
illustrated by Mick Reid

Printed in Mexico

ISBN 978-1-328-77199-5

2 3 4 5 6 7 8 9 10 0908 25 24 23 22 21 20 19 18 17

4500675342 A B C D E F G

My mom asked my brother Max and me to do some shopping for her at the grocery store.

Her list looked strange. I asked my brother about the numbers and letters in parentheses after each item.

Max answered, "They tell us how many units of measurements to buy. You'll learn about these measurements in third grade."

Read•Think•Write Which of these units are used to measure weight?

Our first stop was the deli counter.

Max asked for 5 ounces of cheese, 9 ounces of turkey, and 12 ounces of potato salad.

As the clerk was weighing the food, Max said, "The amounts of these things weigh less than a pound, so people measure their weight in ounces (oz)."

Read•Think•Write What other things would you measure in ounces?

Our next stop was at the produce department.

"Why don't you weigh the fruit?" suggested Max. "We need 2 pounds of apples, 3 pounds of oranges, and 1 pound of grapes."

"What about ounces?" I asked.

"We use pounds (lb) to measure the weight of heavier things," my brother explained.

"There are 16 ounces in 1 pound, so things that weigh more than 16 ounces are usually measured in pounds."

Read·Think·Write How much do the apples weigh?

When we left the produce department, we walked past a display of canned juice.

"How many ounces does this can weigh?" I asked, picking one up.

"Actually," said Max, "for containers of liquids, we talk about capacity, or how much the container holds. The capacity of small containers is usually measured in fluid ounces (fl oz).

"This can holds 8 fluid ounces, or 1 cup (c), of juice."

Read•Think•Write Which is more, 3 cups of water or 16 fluid ounces of water?

The last things on our list were 1 pint (pt) of ice cream, 2 quarts (qt) of milk, and 1 gallon (gal) of water. As we put the items into the shopping cart, Max said, "Pints, quarts, and gallons are used to measure the capacity of containers of different sizes. I remember there are 16 fluid ounces in a pint, and that there are 2 pints in a quart and 4 quarts in a gallon."

Read•Think•Write If 4 children share the pint of ice cream equally, how many fluid ounces will each child get?

As we walked home with our bags, I said to Max, "It was fun shopping with you, and thanks for telling me about units of measurement. But right now I can't wait to get home — these groceries weigh a ton!"

Responding

Vocabulary

1. Put these units of measurement in order from largest to smallest: pint, fluid ounce, quart, cup, gallon.

2. Which unit of measurement does not fit: pint, gallon, pound, fluid ounce? Why?

3. Jessica and Tyler go strawberry picking. When they weigh their baskets, Jessica's scale reads 2 pounds. Tyler's scale reads 32 ounces. Who has more strawberries? (One pound is the same amount as 16 ounces.)

4. Three friends come to Jason's house. Jason divides a 1 quart container of juice into 4 glasses. How much does he pour into each glass? (One quart is the same amount as 32 fluid ounces.)

Activity

Predict Outcomes Find two things in your classroom that you think each weigh about 1 pound. Check each weight by placing the item on a balance. How close were you?